Jupiter

GAS GIANT

Distance to the Sun: 778 million km

Diameter: 142,984 km

Rotation time: 11 years, 315 days

Day length: 9 h, 55 min

Moons: 95 (or more)

Kuiper belt

probably even more asteroids than in the asteroid belt
at least 4 dwarf planets (Pluto, Eris, Makemake, Haumea)

Neptune

GAS GIANT

Distance to the Sun: 4.5 billion km

Diameter: 49,528 km

Rotation time: 165 years

Day length: 15 h, 58 min

Moons: 16 (or more)

Saturn

GAS GIANT

Distance to the Sun: 1.4 billion km

Diameter: 120,536 km

Rotation time: 29 years, 166 days

Day length: 10 h 47 min

Moons: 146 (or more)

Uranus

GAS GIANT

Distance to the Sun: 2.9 billion km

Diameter: 51,118 km

Rotation time: 84 years

Day length: 17 h 14 min

Moons: 28 (or more)

For everyone who works day after day for our little blue dot

First published in the UK in 2024 by Mama Makes Books
www.mamamakesbooks.com
Originally published in Germany in 2021 as
Ein kleiner Blauer Punkt by Magellan GmbH & Co. KG.
© 2021 Magellan GmbH & Co. KG, Bamberg, Germany
English text © Mama Makes Books Ltd.
Illustrated by Maren Hasenjäger. Translated by Leah Francis.
ISBN 978-1-9167800-5-7 (hardback)
ISBN 978-1-9167800-9-5 (paperback)
1 3 5 7 9 10 8 6 4 2
All rights reserved, including the right of reproduction in whole or in part in any form.
A CIP catalogue record of this book is available from the British Library.
Printed in China on FSC paper using eco-friendly inks, glue and varnish.

MAREN HASENJÄGER

Translated by Leah Francis

A little BLUE Dot

Somewhere in the great big universe,

so inconspicuous you could almost miss it,

floats a little blue dot.

If you take a closer look, you'll see that this little blue dot is a planet.

The planet is called Earth – and it's not alone. Together with seven other planets it revolves around a gigantic, bright ball of fire – the Sun.

When Earth sees one of her neighbours passing by again, she lets out a deep sigh. "All the other planets are so special," she says, "I wish I could be like them…"

"Mercury is the closest out of all of us to the Sun. They're probably great friends! Because they are so close to each other, the Sun looks three times bigger from Mercury's surface than it does from mine.

"But you couldn't stand on Mercury without seriously burning your feet! When the Sun is shining there, temperatures can reach over 400 °C. At night, it gets cold and drops down to -170 °C. That's unimaginably cold…"

"And then there's Venus. Aside from the Sun and the Moon, Venus shines the brightest in the sky. She is so bright that people who live on me can see her without a telescope.

"She sparkles so beautifully that many poems have been written about her. Because she's mostly visible at dusk and dawn, she has been called the 'evening' or 'morning star'. What a lovely name…"

The EVENING or MORNING STAR

"But you couldn't go for a walk on Venus either. She is the hottest of all of us planets – it gets even warmer there than on Mercury! The air is poisonous, too, and there are lots of violent thunderstorms."

"There you are, finally!"

"Mars is my next-door neighbour and he's really popular. They keep sending up probes, which are small robots or satellites, to explore him.

"There are lots of mountains on Mars, and one mega-volcano. 'Olympus Mons' is over 20 km high – that's a record-breaking height across all the planets! Wow...

"Mars will probably be the first planet that humans can fly up to visit."

"So far more than 40 probes have been sent to Mars, but not all of them have arrived. Some orbit the planet, while others stay in the spot where they landed. A few have driven around exploring their surroundings, and one little helicopter flew around taking photos!"

Viking 1 (1976–1982)

Sojourner (1997)

Opportunity (2004–2018)

Ingenuity

Perseverance (since 2021)

"Jupiter is the largest of all the planets. He's so massive that I would fit inside him over 1,000 times! But his surface is not solid. It is made of dense, swirling gas. Because he is so big, he pulls lots of comets and asteroids into his orbit – like a gigantic vacuum cleaner.

"A storm has been raging on Jupiter for over 300 years. It is called 'the big red spot' and even that is bigger than me!

"So far, 95 moons have been discovered orbiting Jupiter. On some of them, scientists suspect there might even be huge oceans hidden under kilometres of ice."

Jupiter and me, side by side

Jupiter's big red spot

"Most famous of them all is Saturn. He has beautiful rings that he wears around his stomach like a hula-hoop. That's why I'm really jealous of him...

"But these rings are not a solid loop. If you look closely you can see that they are made up of billions of ice crystals! This was discovered by probes that were sent up to Saturn, such as Cassini-Huygens. They examined not only Saturn's rings, but also the 146 moons that have been found so far. Like Titan, for example, which is even bigger than Mercury, or the snow-white Enceladus, which is made of ice."

Cassini-Huygens (1997–2017)

Enceladus

Titan

Hi!

"Uranus is probably the funniest and most interesting of us all. He's different from the rest of us because he doesn't stand upright. He lies on his side as he rotates.

"That's why the seasons on Uranus are pretty crazy. If you were standing at Uranus's North Pole you would spend 42 years in summer, and the Sun would never set! Afterwards you would have 42 years of winter and eternal night…"

"Then there's Neptune. He's a deep blue, which makes him pretty cool and mysterious. Neptune is the farthest from the Sun, so far away that he can't be seen without a telescope. That's why he was the last of us to be discovered.

"It takes him the longest to make one orbit around the Sun – a full 165 years – but to make up for it, Neptune spins 'galactically' fast. A day on Neptune only lasts 16 hours!

"Perhaps you've heard of Pluto before. He used to be one of the planets in our solar system, until scientists agreed he was far too small for that. Since then he's been counted as one of five dwarf planets, along with Eris, Haumea, Makemake and Ceres."

"And then there's me. Not very big, not particularly warm or cold, no giant storms or eternal summers or beautiful rings. Actually, I'm pretty bor…"

"Now stop that!" interrupts Mars. "I think you're talking nonsense. For instance, we are all jealous of those giant oceans you have. All the water from my surface evaporated, so now I'm bone-dry."

"And here, at the edge of the solar system, it's pretty lonely and cold. I'd much prefer to be as close to the Sun as you are," sighs Neptune.

"Do you have any idea how exhausting it is when you've got 146 moons that won't stop chatting to each other? It must be so nice and quiet when you've only got one moon, like you," interjects Saturn.

Then Earth hears another little voice...

Earth looks around in surprise.

"Look closely, I'm here, really near," calls the voice.

Then Earth sees a child standing in the middle of her waving.

"Finally, you're listening to me," the child says. "I think you've forgotten something very important: you are the only planet I can live on. Here, it's not too hot and not too cold. There is water and food and air to breathe. Only you have grass and crocodiles. And bees and sunflowers. And snow leopards and lakes to swim in and daisies and octopuses and snowball fights and cacti and blue whales and spaghetti and giant redwood trees and children and pizza and bouncy-balls and ants and coral reefs and summer holidays and ice cream with whipped cream on the top and much, much more!"
The child pauses and then smiles:

"You are exactly right for me!"

At that, Earth blushes and smiles...

so inconspicuous you could almost miss it,

floats a little blue dot.

A very special dot.

Sun

Distance to the nearest star (Alpha Centauri): 4.34 light years

Diameter: 1,392,700 km

Rotation time: 25 days

Mars

ROCKY PLANET

Distance to the Sun: 228 million km

Diameter: 6,790 km

Rotation time: 687 days

Day length: 1 day, 37 min

Moons: 2

Venus

ROCKY PLANET

Distance to the Sun: 108 million km

Diameter: 12,100 km

Rotation time: 225 days

Day length: 243 days

Moons: 0

Earth

ROCKY PLANET

Distance to the Sun: 150 million km

Diameter: 12,735 km

Rotation time: 365 days

Day length: 23 h 56 min

Moons: 1

Mercury

ROCKY PLANET

Distance to the Sun: 58 million km

Diameter: 4,880 km

Rotation time: 88 days

Day length: 58 days, 15 h 36 min

Moons: 0

Asteroid belt

over 650,000 asteroids

at least 1 dwarf planet (Ceres)

Distances and times are approximate